Optogenetic Technologies and Applications Conference 2019

Boston, Massachusetts, USA
8 – 10 December 2019

ISBN: 978-1-7138-0570-0

Printed from e-media with permission by:

Curran Associates, Inc.
57 Morehouse Lane
Red Hook, NY 12571

Some format issues inherent in the e-media version may also appear in this print version.

Printed with permission by Curran Associates, Inc. (2020)

For permission requests, please contact AIChE
at the address below.

AIChE
120 Wall Street, FL 23
New York, NY 10005-4020

Phone: (800) 242-4363
Fax: (203) 775-5177

www.aiche.org

Additional copies of this publication are available from:

Curran Associates, Inc.
57 Morehouse Lane
Red Hook, NY 12571 USA
Phone: 845-758-0400
Fax: 845-758-2633
Email: curran@proceedings.com
Web: www.proceedings.com

TABLE OF CONTENTS

OPTOGENETICS IN BIOMEDICINE - CARDIOVASCULAR MEDICINE

OPTOGENETICS IN BIOMEDICINE - DRUG DELIVERY

OPTOGENETICS IN BIOMEDICINE – NEUROBIOLOGY

OPTOGENETICS IN BIOMEDICINE - OTHER BIOMEDICAL APPLICATIONS

OPTOGENETICS IN BIOTECHNOLOGY – MICROBES

OPTOGENETICS IN BIOTECHNOLOGY – PLANTS

Bidirectional Optogenetic Control of Neurotrophin Signaling during Cell Differentiation and Embryonic Development

Conference: Optogenetic Technologies and Applications

Year: 2019

Proceeding: Optogenetic Technologies and Applications

Group: General Submissions

Session: Advances in Optogenetic Technologies - Advanced optogenetic technologies for animal cells

Authors:

Zhang, K., University of Illinois

The neurotrophin signaling pathway regulates a wide spectrum of cellular functions such as cell survival, proliferation, differentiation, and apoptosis. It also plays a key role in cell fate determination during embryonic development. Evidence suggests that the signaling output of the neurotrophin pathway varies with its temporal kinetics. However, a quantitative delineation of signaling kinetics is limited due to a lack of tools that allows precise control of the neurotrophin signaling in time and space. Non-neuronal optogenetics, an emerging technology that utilizes light to control intracellular signaling pathways, offer an alternative solution to address this challenge. In this presentation, I will introduce optogenetic systems recently developed in our laboratory that allow for reversible and bidirectional optical control of neurotrophin signaling pathway in intact cells and in developing *Xenopus laevis* embryos. I will also discuss the limitations of current non-neuronal optogenetics and update you with current progress in the field in overcoming these limitations.

Extracellular Optogenetics to Control Cell Fate and Function

Conference: Optogenetic Technologies and Applications

Year: 2019

Proceeding: Optogenetic Technologies and Applications

Group: General Submissions

Session: Advances in Optogenetic Technologies - Advanced optogenetic technologies for animal cells

Authors:

Weber, W., SGBM Spemann Graduate School of Biology and Medicine, University of Freiburg

Optogenetics is impacting life science research and development in waves of innovation. With the introduction of light-sensitive ion channels into neuronal cells, the first wave of optogenetics enables fascinating insight and innovation in neurobiology. A similar wave occurred shortly after with the use of light-responsive proteins and enzymes to control signaling events at every level of the signaling cascade, from membrane receptors via kinases to gene expression and protein degradation (seewww.optobase.orgfor an overview). Currently, a third wave is being triggered with the introduction of extracellular optogenetics, where purified photoreceptors are engineered and used in the extracellular space. In this context we will demonstrate efficient protocols for the gram-scale production of photoreceptor proteins and their application in the extracellularly mediated control and analysis of cell signaling. Examples include synthetic extracellular matrices with light-responsive reversibly adjustable properties, the release and capture of extracellular biomolecules, optically controlled affinity chromatography for interaction proteomics or the design of information-processing biohybrid materials.

Near-Infrared Light-Activated Adenylate Cyclase Adapted for Applications in Mammals

Optogenetic
Technologies & Applications

Conference: Optogenetic Technologies and Applications

Year: 2019

Proceeding: Optogenetic Technologies and Applications

Group: General Submissions

Session: Advances in Optogenetic Technologies - Advanced optogenetic technologies for animal cells

Authors:

Gomelsky, M., University of Wyoming
Fomicheva, A., University of Wyoming
Zhou, C., University of Wyoming
Sun, Q. Q., University of Wyoming

Light in the near-infrared optical window (NIRW) penetrates deep through mammalian tissues, including the skull and brain tissue. We engineered an adenylate cyclase (AC) activated by NIRW light (NIRW-AC) and adapted for mammalian optogenetic applications. A series of fusions of the bacteriophytochrome photosensory modules and homodimerc bacterial AC modules were engineered using previously deciphered guidelines for designing homodimeric bacteriophytochromes. One fusion, designated IlaM5, was found to have higher activity at 37 °C and be better expressed in mammalian cells. In favorable contrast to the NIRW-ACs engineered earlier, it can mediate cAMP-dependent photoactivation of gene expression in mammalian cells. The *ilaM5* gene was cloned in an AAV vector and delivered into the ventral basal thalamus region of the mouse brain resulting in the photosuppression of the cAMP-dependent wave pattern of the sleeping brain known as spindle oscillations. Robust and reversible spindle oscillation suppression in sleeping mice can be achieved by exposing mice to NIRW light from an

LED panel or a hand-held laser. This study supports validity of engineering principles for homodimeric bacteriophytochrome proteins, describes a NIRW-AC suitable for mammalian applications, and demonstrates feasibility of bacteriophytochrome-mediated control of the brain activity via transcranial irradiation.

Near-infrared optogenetic tools engineered from bacterial phytochromes

Conference: Optogenetic Technologies and Applications

Year: 2019

Proceeding: Optogenetic Technologies and Applications

Group: General Submissions

Session: Advances in Optogenetic Technologies - Advanced optogenetic technologies for animal cells

Authors:

Verkhusha, V., Albert Einstein College of Medicine

Near-infrared light is favorable for use in mammalian tissues due to its low scattering and low absorbance by hemoglobin, melanin and water. Therefore, optogenetic tools for light manipulation in mammals should have action spectra within the near-infrared tissue transparency window. Interestingly, natural bacterial phytochromes utilize the low molecular weight biliverdin, found in most mammalian tissues, as a photoreactive chromophore. Due to their near-infrared absorbance and availability of endogenous biliverdin chromophore, bacterial phytochromes are preferred templates for designing of optogenetic tools for noninvasive light-control of biochemistry and physiology of living animals. Moreover, bacterial phytochromes spectrally complement existing genetically encoded fluorescent probes and optogenetic constructs that operate in the visible spectral range. Several bacterial phytochromes were already developed into the near-infrared optogenetic tools. The design strategies, properties *in vitro* and in mammalian cells, and applications of these tools are discussed.

Comparative Studies of Lov Switching Mechanisms and Characteristics

Optogenetic Technologies & Applications

Conference: Optogenetic Technologies and Applications

Year: 2019

Proceeding: Optogenetic Technologies and Applications

Group: General Submissions

Session: Advances in Optogenetic Technologies - Advanced optogenetic technologies for microbial and plant cells

Authors:

Gardner, K. H., City College of New York

Environmental cues regulate many biological processes, coordinating cellular pathways to respond to changing conditions. Such regulation is often initiated by sensory protein domains which expand their chemical repertoire by using small molecule ligands to convert environmentally-triggered changes into altered protein/protein interactions. Light-Oxygen-Voltage (LOV) domains present outstanding model systems for such processes, given their abilities to convert blue light into photochemically-driven formations of protein/flavin adducts and subsequent triggering of the allosteric control of effectors as diverse as kinases to DNA-binding domains. Using a combination of biophysics, biochemistry and cell biology, we seek to gain insight into the mechanistic controls of such environmental sensing domains for both fundamental understanding and subsequent artificial control.

To address these questions, we combine approaches from: 1). biophysics, including NMR and EPR spectroscopies plus HDX-MS to monitor conformation and dynamics, 2). *In vitro* biochemistry studies of function, and 3). Cell-based examination of function in complex settings. Integrated data from these methods across LOV-HTH, LOV-HK and RGS-LOV-DUF settings will be discussed. This information has shown several commonalities of LOV signaling in diverse classes of sensory proteins, and further, develop structure-based mutations to shift these proteins among these states. In particular, we note substantial differences in the degree of conformational rearrangements observed among photoactivation among different LOV proteins. Further, we also observe a range of difference in â€œoffâ€ kinetics among proteins under study, often with substantial differences in the timing of adduct breakage and functional deactivation. We have taken advantage of this mechanistic understanding to develop the artificial regulation of such systems, both *in vitro* and in living cells.

Taken together, our work integrates information across a fascinating class of natural switches and suggests routes by which these can be manipulated to achieve a wide range of technical outcomes.

Controlling Nucleic Acids By Light

Optogenetic Technologies & Applications

Conference: Optogenetic Technologies and Applications

Year: 2019

Proceeding: Optogenetic Technologies and Applications

Group: General Submissions

Session: Advances in Optogenetic Technologies - Advanced optogenetic technologies for microbial and plant cells

Authors:

M□glich, A., University of Bayreuth

Controlling Nucleic Acids by Light

Sensory photoreceptors underpin diverse adaptations of organismal behavior, lifestyle and physiology to incident light. In optogenetics, photoreceptors double as genetically encoded, light-gated actuators and enable the noninvasive control of cellular circuits with spatiotemporal precision. Against this backdrop, we investigate and engineer blue-light-responsive receptors of the light-oxygen-voltage (LOV) family. Certain of these photoreceptors unlock novel optogenetic modalities, in particular the control of nucleic-acid-based processes, e.g., transcription, translation and endonuclease activity. Biochemical analyses of receptor structure, function and signaling mechanism unravel the molecular bases for light-dependent allostery and inform additional protein engineering efforts.

Engineering photoreceptors into optogenetic tools for the controland understanding of cellular processes in animal and plantsystems

Optogenetic
Technologies & Applications

Conference: Optogenetic Technologies and Applications

Year: 2019

Proceeding: Optogenetic Technologies and Applications

Group: General Submissions

Session: Advances in Optogenetic Technologies - Advanced optogenetic technologies for microbial and plant cells

Authors:

Zurbriggen, M., University of Freiburg

Our synthetic biology research focuses on engineering bacterial and plant photoreceptors sensitive to different wavelengths of the white light spectrum (UV-B, blue, green, red/far-red) into synthetic photoswitches rewired to control molecular processes with high precision, quantitative and high spatio-temporal resolution, in a non-invasive way and with minimized toxicity. We implement these molecular tools into microbial, mammalian and plant cells, and in vivo in animals and plants for selectively manipulating signaling networks and metabolic pathways. This synthetic biology approach opens up unforeseen perspectives in fundamental and applied research, as exemplified hereby in the study of signalling pathways, including those involving plant photoreceptors, and crop design.

Inducible Asymmetric Cell Division and Cell Differentiation in a Bacteria

Conference: Optogenetic Technologies and Applications

Year: 2019

Proceeding: Optogenetic Technologies and Applications

Group: General Submissions

Session: Advances in Optogenetic Technologies - Advanced optogenetic technologies for microbial and plant cells

Authors:

Mushnikov, N., University of Wyoming

Multicellular organisms achieve greater complexity through cell divisions that generate different cell types. We engineered a simple genetic circuit that induces asymmetric cell division and subsequent cell differentiation in Escherichia coli. The circuit involves a scaffolding protein, PopZ, that is stably maintained at a single cell pole over multiple asymmetric cell divisions. PopZ was functionalized to degrade the signaling molecule, c-di-GMP. By regulating synthesis of functionalized PopZ via small molecules or light, we can chemically or optogenetically control the relative abundance of two distinct cell types, characterized by either low or high c-di-GMP levels. Differences in c-di-GMP levels can be transformed into genetically programmable differences in protein complex assembly or gene expression, which in turn produce differential behavior or biosynthetic activities. Programmable bacterial cell differentiation is a new genetic tool of synthetic biology and biotechnology. We called this strategy as Microbial Stem Cell Technology (MiST), which enables a growing microbial culture to stably maintain two or more distinct cell types in a ratio that can be genetically programmed and/or dynamically controlled during cultivation. It is contemplated that employing of MiST would increase product yield in microbial fermentations and advanced engineering of biomaterials.

Engineering K+channels for Long Term Inhibition

Optogenetic
Technologies & Applications

Conference: Optogenetic Technologies and Applications

Year: 2019

Proceeding: Optogenetic Technologies and Applications

Group: General Submissions

Session: Advances in Optogenetic Technologies - Novel optogenetic modalities

Authors:

Moroni, A., University of Milano
Saponaro, A., University of Milan
Laskowski, M., University of Milan
Engel, A., TU-Darmstadt
Thiel, G., Technische Universit□t-Darmstadt

We are using different strategies for modulating potassium (K^+) channel activity by light. One strategy relies on the engineering of a fully genetically-encoded light regulated K^+channel (BLINK), obtained by fusing LOV2 domain to the small viral channel Kcv. We have optimized this system in such a way that it expresses in neurons and allows long lasting light-off activity and, in turn, long lasting inhibition of neurons. A second approach is based on light-dependent modulation of native HCN channels. These channels that control pacemaking activity in brain and heart, have a regulatory subunit TRIP8b that prevents cAMP induced enhancement of their current. We have reduced TRIP8b to a short peptide, nanoTRIP, and fused it to a LOV2 domain that releases the active peptide upon blue light. A final strategy uses a light-dependent dimerization system to activate the transcription of small K^+channels. With this system we can target active K+ channels not only to the plasma membrane but also to the inner mithochondrial membrane. This approach allows long lasting depolarization of the mitochondria with physiological consequences, such as modulation of mitochondrial calcium content.

On-Demand Targeted Light Generation in Bio-Compatible Elastomers Using High-Intensity Focused Ultrasound for Optogenetic Application

Optogenetic
Technologies & Applications

Conference: Optogenetic Technologies and Applications

Year: 2019

Proceeding: Optogenetic Technologies and Applications

Group: General Submissions

Session: Advances in Optogenetic Technologies - Novel optogenetic modalities

Authors:

Kim, G., University of Illinois at Urbana Urbana-Champaign
Chu, J., Beckman Institute, University of Illinois at Urbana-Champaign
Wu, Q., University of Illinois at Urbana?Champaign
Halmes, A., University of Illinois at Urbana?Champaign
Oelze, M., University of Illinois at Urbana?Champaign
Moore, J., University of Illinois at Urbana?Champaign
Li, K., University of Illinois at Urbana?Champaign

Currently most optogenetic activations are triggered by LED that usually requires invasive optical fiber implants, which presents challenges for in vivo studies and biomedical applications. Herein we introduce a new targeted, remote light generation platform based on high-intensity focused ultrasound (HIFU) that exploits polymer mechanochemistry to spatiotemporally control the remote emission of light (1). As a proof of concept, 1.5 wt% of covalently incorporated 1,2-dioxetane mechanophore was embedded as a crosslinker in a 5 mm thick PDMS film. Upon scission of the dioxetane, the mechanophore generated an excited state ketone and in the process emitted luminescence. The luminescence was enhanced by an energy acceptor, 9,10-

diphenylanthracene (DPA). In a tank of degassed water, a HIFU transducer (550 kHz) was employed to generate continuous wave ultrasound for 5 seconds that triggered the prepared dioxetaneâ€“functionalized PDMS film. Blue luminescence (ca. 470 nm) was observed from the PDMS film through remote, non-invasive control via HIFU. However, the measured intensity (ca. 0.2 Î¼W/cm^2) was weaker than the reported threshold (ca. 1~10 mW/cm^2) for optogenetic applications (2). This low intensity is due to the weakly-dispersible DPA and can be overcome by adopting a more soluble hexyl-functionalized DPA (HDPA). Incorporating 7.5 wt% HDPA into PDMS improved the solubility of the energy acceptor and the emitted light intensity by two orders of magnitude. We believe this increased intensity could be promising for optogenetic applications. This study explores HIFU as a potential stimulus that is capable of triggering on-demand, spatiotemporally resolved mechanoluminescent transduction to leverage the advantages of existing optogenetic technologies.

References

1. Kim G, et al. (2019) High-intensity focused ultrasound-induced mechanochemical transduction in synthetic elastomers. *Proc Natl Acad Sci* 116(21):10214â€“10222.

2. Raman R, et al. (2016) Optogenetic skeletal muscle-powered adaptive biological machines. *Proc Natl Acad Sci* 113(13):3497â€“3502.

Rhodopsin-cyclases for photocontrol of cGMP/cAMP signaling

Optogenetic
Technologies & Applications

Conference: Optogenetic Technologies and Applications

Year: 2019

Proceeding: Optogenetic Technologies and Applications

Group: General Submissions

Session: Advances in Optogenetic Technologies - Novel optogenetic modalities

Authors:

Hegemann, P., Humboldt University

The recently discovered photoreceptor family of enzyme rhodopsins comprises a number of Rhodopsin-cyclases (RhGC/ACs) and Rhodopsin-phosphodiesterases (RhPDEs). These are rhodopsins that are either directly connected to an enzyme as realized in fungal class of Blasticladiomycetes (Avelar et al. 2014) or the coupling is indirect via a His-kinase and a Response Regulator as in case of the Histidine Kinase Rhodopsins of Chlorophyte algae (Luck et al. 2012).

We characterized two recombinant Rhodopsin-cyclases from*Blastocladialla emersonii* (BeRhGC) (Scheib et al. 2015) and *Catenaria anguillulae* (CaRhGC) (Scheib et al. 2018) and have shown that both enzymes are completely inactive in the dark and are active upon illumination. The enzymatically active state is formed within 10 to 40 ms after a flash and decays within 100 to 500 ms depending on the species. The cyclases are highly selective for GTP whereas ATP is acting as a competitive inhibitor. Coexpression of RhGC with a cGMP-channel from olfactory neurons and other sources in Xenopus oocytes , ND cells, HEK cells or neurons generating a large photocurrent upon illumination and conferring a high light sensitivity to the cell of interest. We are currently working on a cGMP-sensitive potassium channel in order to use our Two-Component Approach for rhodopsin-based hyperpolarization in the neurosciences and cell biology.

In parallel we work on Chlamydomonas Histidine kinase rhodopsins (HKRs) from *Chlamydomonas reinhardtii* (Luck and Hegemann 2017; Luck et al. 2012)and *Ostreococcus tauri* (Scheib unpublished). Both are synthesized in the cell as blue-green sensitive rhodopsins but are converted upon illumination into a UVA-sensitive species (Rh-UV) with maximal absorption near 380 nm. The thermal stability of the Rh-UV is very high but varies between many minutes to hours but it can be photochemically back-converted upon illumination with UVA light.

Soft, Wireless Optoelectronic Systems for Optogenetics

Optogenetic
Technologies & Applications

Conference: Optogenetic Technologies and Applications

Year: 2019

Proceeding: Optogenetic Technologies and Applications

Group: General Submissions

Session: Advances in Optogenetic Technologies - Novel optogenetic modalities

Authors:

Rogers, J. A., Northwestern University

Advanced optoelectronic systems that intimately integrate with the brain and the peripheral nervous system have the potential to accelerate progress in neuroscience research and to spawn new therapies in clinical medicine. Specifically, capabilities for injecting electronics, light sources, photodetectors, multiplexed sensors, programmable microfluidic networks and other components into precise locations of the deep brain and for softly laminating them onto targeted regions of the surfaces of the brain or the peripheral nerves will open up unique and important opportunities in stimulation, inhibition and real-time monitoring of neural circuits via the combined techniques of optogenetics, photometry, electrical stimulation/sensing and local drug delivery. In this talk, we will describe foundational concepts in materials science and assembly processes for these types of technologies. Examples in system level demonstrations include experiments on freely moving animals with â€˜cellular-scaleâ€™, injectable optofluidic neural probes for optogenetics research and with closed-loop systems for optogenetic control of bladder function.

Synthesis and Luminescence Investigation of Nano-Scintillators As Next Generation Tools for Optogenetics

Conference: Optogenetic Technologies and Applications

Year: 2019

Proceeding: Optogenetic Technologies and Applications

Group: General Submissions

Session: Advances in Optogenetic Technologies - Novel optogenetic modalities

Authors:

Dickey, A., Clemson University
Zhang, E., Clemson University
Fougler, S. H., Clemson University
Kolis, J. W., Clemson University

Currently, the visible light sources used for optogenetics are surgically implanted into the brain tissue, but this harmful and invasive technique may be avoided if suitable scintillating nanoparticles can be inserted via injection. The proposed nano-scintillator particles must adhere to rigorous parameters including being under 100nm, uniform, nontoxic, and dispersible to be successful in this biological system. The phase, crystallinity, and dopant concentration must be optimized to absorb X-ray radiation and emit photons of the desired energy necessary to activate the selected neurons.

We began a synthesis program to prepare suitable cerium doped gadolinium silicate nanoparticles to pair with blue absorbing opsins, such as channelrhodopsin-2. This produced several phases including Ce:$Gd_{4.67}(SiO_4)_3O$ (Ce:GSAP), Ce:Gd_2SiO_5 (Ce:GSO), and

$Ce:Gd_2Si_2O_7$, (Ce:GPS) nano-scintillators. In our modified core-shell synthesis the silica core served as both a template to ensure uniformity in size and shape, and could be manipulated to elicit different silicate phases. The addition of lutetium into the gadolinium silicate nanoparticles was also investigated. The mixed gadolinium/lutetium materials could also be engineered to produce different silicate phases and showed enhanced light output under X-ray stimulation.

Next, europium doped rare earth oxide ($Eu:RE_2O_3$, RE = Y, Gd, Lu) nanospheres were targeted as light sources for inhibitory opsins, such as the red-shifted cruxhalorhodopsin, Jaws. Monodispersed, spherical RE_2O_3 nanoparticles were produced via an urea assisted homogenous precipitation followed by annealing. We were able to obtain systematic increases in the radioluminescence intensity from our base $Eu:RE_2O_3$ nanoparticles by applying two different synthetic strategies.

MTS assays were ran on human embryonic kidney (HEK) cells incubated with the nanoparticles as an early cytotoxicity assessment. The ability to produce these particles in a straightforwardly, uniform, and dispersible manner as well as their relative chemical and biological inertness make them exciting candidates for the optogenetics toolbelt.

Multiplexed, Multicolor Optogenetics for High-Throughput Manipulation of Cell Behavior

Conference: Optogenetic Technologies and Applications

Year: 2019

Proceeding: Optogenetic Technologies and Applications

Group: General Submissions

Session: Advances in Optogenetic Technologies: Materials and Devices for Optogenetic Applications

Authors:

Bugaj, L., University of California, San Francisco
Lim, W., University of California, San Francisco
Xiong, X., University of California, San Francisco

Optogenetic control of biomolecular reactions can, in principle, allow data-rich experiments with precise perturbations and multiple channels of control (multiple stimulation colors). Moreover, because light patterns are programmable and light can be applied remotely, optogenetic experiments could be readily multiplexed for high-throughput manipulation without the need for expensive robotics, bespoke microfluidics platforms, and perturbative media exchanges. However, there is a lack of hardware to perform optogenetic stimulation in a high-throughput microwell format. Here, we describe the optoPlate-96, an LED-based device for independent control of 96 3-color optogenetic experiments in 96- or 384-well plates. The optoPlate-96 is compatible with long-term illumination both in cell culture incubators, as well as in live-cell microscopy. We will first describe the optoPlates and show how they enable robust and reproducible optogenetic experiments. Then, we will detail two studies that highlight the utility of optoPlate experiments. In the first, we used the optoPlate to systematically assess how cancer cells process Ras-Erk signal dynamics differently from normal cells. In the second study, we demonstrated how 3-color (blue, red, far-red) optoPlate-96 illumination enables orthogonal control of red- and blue-responsive optogenetic proteins within the same cell. We used multicolor probing to study signal integration between Ras- and PI3K signaling and uncovered new synergies between these two well-studied pathways. We anticipate that the optoPlate-96 will find broad utility and will help realize the potential of optogenetics for the quantitative dissection and manipulation of living systems.

Opto Biolabs – Combining Optogenetics with Flow Cytometry

Conference: Optogenetic Technologies and Applications

Year: 2019

Proceeding: Optogenetic Technologies and Applications

Group: General Submissions

Session: Advances in Optogenetic Technologies: Materials and Devices for Optogenetic Applications

Authors:

Brenker, K., Opto Biolabs / University of Freiburg

Optogenetic tools allow the remote control of cellular pathways using light. This enables the isolated, functional investigation of almost any signaling molecule within complex signaling pathways. More interestingly, the remote control of cellular pathways has major implications for therapeutic tools, including targeted immunotherapy.

A major obstacle for optogenetic researchers is the controlled delivery of light to the cell sample and hence the most popular tools for optogenetic studies are microscopy-based cell analyses and *in vitro* experiments. The flow cytometer has major advantages over a microscope, including the ability to rapidly measure thousands of cells at single cell resolution. However, it is not yet widely used in optogenetics. Opto Biolabs, a spinoff from the University of Freiburg, develops specialized illumination devices enabling the combination of optogenetics and flow cytometry.

Optogenetic flow cytometry can be applied for many optically controlled substances including cage compounds or photoconvertible fluorescent proteins and substantially expands the set of possible experiments. Here, we present first results from our research collaboratoins using this novel technology.

Our future work will focus on the expansion of our technology to optogenetic cell sorting, which will greatly simplify the discovery and development of novel, cell-based optogenetic therapies.

Genetically Encoded Voltage Imaging in Induced Pluripotent Stem Cell Cardiomyocytes

Optogenetic Technologies & Applications

Conference: Optogenetic Technologies and Applications

Year: 2019

Proceeding: Optogenetic Technologies and Applications

Group: General Submissions

Session: Optogenetics in Biomedicine - Cardiovascular medicine

Authors:

Kim, B., Stanford University
Lin, M., Stanford University
Wu, H., Stanford University
Wu, J., Stanford University

Characterization of induced pluripotent stem cell derived cardiomyocytes (iPSC-CMs) usually requires looking for beating frequency, membrane potential changes, shape, and movement. The gold standard for characterizing iPSC-CMs in cell culture is patch-clamp electrophysiology, but this method is low-throughput and destructive. An alternative method may be to use fluorescent voltage indicators. Preferentially, fluorescent voltage indicators would be ratiometric so that changes in membrane potential can be discriminated from motion-induced changes in fluorescence. Here we report a ratiometric variant of the voltage indicator ASAP3 in which the green fluorescent ASAP3 is fused to a new cyan-excitable fluorescent protein, mCyRFP2. ASAP3-mCyRFP2 enables ratiometric voltage imaging using single excitation dual emission microscopy. Additionally, we demonstrate we can faithfully report drug-induced fluorescence dependent voltage changes by adding isoproterenol. Finally, we show that upon movement of a beating cell, the change in fluorescence ratios can correct for motion artifacts.

Optogenetics to Explore G Protein Signaling in the Heart

Optogenetic
Technologies & Applications

Conference: Optogenetic Technologies and Applications

Year: 2019

Proceeding: Optogenetic Technologies and Applications

Group: General Submissions

Session: Optogenetics in Biomedicine - Cardiovascular medicine

Authors:

Bruegmann, T., University Medical Center Goettingen

Besides manipulating the membrane potential by light induced de- and hyperpolarization, optogenetic G protein coupled receptors offer a unique potential to investigate the activation of the specific intracellular signaling cascades. In contrast to pharmacological stimulation, this new approach provides unprecedented spatial and temporal control which is especially intriguing looking into the specific effects within different regions of the intact heart. Furthermore, optogenetic approaches allow for the first time to determine the kinetics of the single steps within the signaling cascades and the different microdomaines. This is of special interest as it allows to investigate when a primarily physiological effect turns into a pathophysiological one.

Therefore this talk will give an overview about recent progress in this new branche within cardiac optogenetics and its new proteins development to specifically controll the Gi, Gs and Gq pathway. In addition, I will present different possible application to investigate cardiac physiology and pathophysiology.

Optogenetics-Enabled High-Throughput Technology for the Advancement of Personalized Medicine with in Human Ips-Cardiomyocytes

Optogenetic Technologies & Applications

Conference: Optogenetic Technologies and Applications

Year: 2019

Proceeding: Optogenetic Technologies and Applications

Group: General Submissions

Session: Optogenetics in Biomedicine - Cardiovascular medicine

Authors:

Entcheva, E., The George Washington University

Optical targeting (stimulation or recording) allows distributed parallel access to thousands and even millions of cells and locations at the same time, and within the tissue setting; optical targeting is high-throughput by nature. The combination of optogenetic stimulation with optical imaging of electrical function in cardiomyocytes, i.e. the realization of â€œall-optical electrophysiologyâ€ in a high-throughput (HT) manner is essential for the rapid advancement of personalized medicine using human stem-cell-derived cardiomyocytes (iPS-CMs). It represents one of the short-term achievable impact applications of optogenetics as all elements are ready for immediate translation. We have developed and optimized technology for automated acute and chronic monitoring of human iPS-CMs that enables high-quality high-content characterization of such patient-derived cells for disease modeling, drug screening and mechanistic studies of these heart cells and engineered tissues. Such approaches also can help in improving experimental design, removing bias, increasing reproducibility and strengthening predictive power for faster translation of basic science findings.

Engineering of Far-Red Light-Controlled Optogenetic Devices for Biomedical Applications

Optogenetic
Technologies & Applications

Conference: Optogenetic Technologies and Applications

Year: 2019

Proceeding: Optogenetic Technologies and Applications

Group: General Submissions

Session: Optogenetics in Biomedicine - Drug Delivery

Authors:

Ye, H., East China Normal University

Synthetic biology-inspired gene- and cell-based therapies are recognized as central pillars of next-generation medicine. However, the controllability remains a critical issue for clinical applications. Optogenetics is opening up exciting new opportunities for precision-guided medicine by illumination with light as a trigger signal to achieve spatiotemporal control of cellular activities. Optogenetics-based technology promises the capability to achieve traceless, remotely controlled precision dosing of an enormous range of therapeutic outputs. In this talk, I will introduce a far-red light (FRL)-controlled transgene expression device that was created for diabetes therapy. Using a multidisciplinary design principle coupling electrical engineering, software development, and synthetic biology, we have engineered a technological infrastructure enabling the smartphone-assisted semiautomatic treatment of diabetes in mice. By combining electronic device-generated digital signals with optogenetically engineered cells, this study provides a step toward translating cell-based therapies into the clinic. Furthermore, the FRL-controlled transgene expression device was further combined with CRISPR-Cas9/dCas9 tool to obtain a FRL-Cas9 device for controllable gene editing and a FRL-activated CRISPR-dCas9 effector (FACE) system for epigenome remodeling. The FRL-Cas9 device enables gene editing both in mammalian cells and animals controlled by FRL illumination. The FACE system enables transcriptional activation of user-defined endogenous genes under FRL illumination with high induction efficiency (~500-fold), and demonstrated to mediate targeted epigenetic modulation in mammalian cells and animals. Optogenetic devices developed in this work expanded the optogenetic toolkit for the development of therapeutic interventions and precise genome engineering in many areas of basic and translational research, which may in turn boost the clinical progress of optogenetics-based precision medicines.

Optogenetic Engineering of Pancreatic Beta-Cells for Modulation of Insulin Release By Light

Conference: Optogenetic Technologies and Applications

Year: 2019

Proceeding: Optogenetic Technologies and Applications

Group: General Submissions

Session: Optogenetics in Biomedicine - Drug Delivery

Authors:

Tzanakakis, E. S., Tufts University
Zhang, F., Tufts University

Diabetes is characterized by abnormal blood sugar levels linked to damage or ablation of pancreatic beta-cells due to autoimmunity (type 1 diabetes; T1D) or insulin resistance (type 2 diabetes; T2D). Over 30 million people in the US suffer from diabetes. Insulin administration is essential for the treatment of all T1D patients and those with late stage T2D. Given the limited supply of donor islet cells, an appealing alternative is the engineering of beta-cells aiming at increasing the specific rate glucose-stimulated insulin secretion (GSIS) thereby reducing the total number of cells needed for reconstituting normoglycemia. In the present study, we implemented optogenetic engineering for enhancing GSIS in beta-cells via photostimulation rather than pharmacological agents. This was accomplished by ectopic expression of a photoactivated adenylyl cyclase (PAC) gene in beta-cells followed by extensive testing of their function *in vitro* and *in vivo*. Rodent Î^2-cell lines and primary islet cells expressing PAC and exposed to blue light exhibited increased intracellular cyclic adenosine monophosphate (cAMP) and GSIS rate compared to their counterparts without PAC expression and/or photostimulation. Stable PAC expression and successive rounds of stimulation with light had no adverse effect on cell growth and viability over multiple passages. Cyclic AMP was also elevated with light in the absence of glucose but insulin secretion did not increase in agreement with the role of cAMP as an amplifier â€“ not an inducer â€“ of GSIS and the preservation of GSIS dependence on glucose. Encapsulated PIs were implanted in streptozotocin-treated diabetic mice. Animals receiving engineered cells displayed improved glucose tolerance and higher plasma insulin with light vs. those kept in dark. These findings support further development of optogenetic augmentation of beta-cell GSIS to enable cell therapy technologies for diabetes and tools for studying beta-cell physiology in normal and disease states.

All-Optical Electrophysiology for Disease Modeling and Drug Discovery

Optogenetic
Technologies & Applications

Conference: Optogenetic Technologies and Applications

Year: 2019

Proceeding: Optogenetic Technologies and Applications

Group: General Submissions

Session: Optogenetics in Biomedicine - Neurobiology

Authors:

Cohen, A. E., Harvard University

The electrical activity of neurons is fundamental to their function, and dysregulation of activity is associated with many nervous system disorders. However, measurement of membrane potential has historically been difficult. We developed tools for all-optical electrophysiology: simultaneous optogenetic stimulation and voltage imaging in large ensembles of neurons. I will describe how these tools are being used for drug discovery and development, in systems ranging from sensory neurons to human stem cell-based models of neurodevelopmental and neurodegenerative diseases.

Biological Light for Optogenetic Applications in Biomedicine

Optogenetic
Technologies & Applications

Conference: Optogenetic Technologies and Applications

Year: 2019

Proceeding: Optogenetic Technologies and Applications

Group: General Submissions

Session: Optogenetics in Biomedicine - Neurobiology

Authors:

Hochgeschwender, U., Central Michigan University

Biological light, or bioluminescence, is generated when a luciferase enzyme oxidizes a luciferin, a small molecule substrate. The emitted light enables activation of nearby photoreceptors, thus allowing to drive optogenetic elements from an internal â€œcold lightâ€ source when the substrate is administered. Advantages are the avoidance of implanted light fibers, the absence of heat damage, and the ability to reach every cell that expresses the luciferase-opsin combination rather than only those in the vicinity of a physical light source. We took advantage of this bioluminescent optogenetics approach in two preclinical applications that pose specific challenges to conventional optogenetics. In a spinal cord injury model in rats we genetically expressed a luminopsin (LMO3), a fusion protein of the light emitting Gaussia luciferase tethered to the light sensing Volvox channelrhodopsin 1, in motorneurons and interneurons below the site of injury (1). Neurons were activated by the addition of the luciferase substrate coelenterazine (CTZ), thus allowing non-invasive stimulation and recruitment of all targeted neurons. Neural stimulation starting a day after contusion injury of the thoracic spine for two weeks of either interneuron or motorneuron populations below the level of injury significantly improved locomotor recovery lasting beyond the time of stimulation. In a mouse Parkinsonâ€™s model we utilized bioluminescent optogenetics to activate transplanted neural precursor cells (2).

Remote stimulation of all transplanted cells expressing LMO3 by daily application of CTZ improved motor behavior in Parkinsonâ€™s disease mice. This bioluminescent optogenetics approach might be beneficial across a wide range of stem cell transplantation therapies for optimal neural repair and functional recovery.

1. Petersen et al., Restoring function after severe spinal cord injury through bioluminescence-driven optogenetics. bioRxiv 710194, 2019
2. Zenchak et al., Bioluminescence-driven optogenetic activation of transplanted neural precursor cells improves motor deficits in a Parkinson's disease mouse model. J Neurosci Res. Mar 25, 2018

Independent Two-Colour Control of Neuronal Activity

Optogenetic
Technologies & Applications

Conference: Optogenetic Technologies and Applications

Year: 2019

Proceeding: Optogenetic Technologies and Applications

Group: General Submissions

Session: Optogenetics in Biomedicine - Neurobiology

Authors:

Gee, C. E., University Medical Center Hamburg-Eppendorf

Synaptic plasticity, the activity-dependent strengthening or weakening of the connections (synapses) between coincidently active neurons in the brain, underlies learning and memory. Repetitive pairing of activity (spiking) leads to synaptic potentiation when the presynaptic input is causal (i.e. precedes spiking of the postsynaptic neuron) or to depression when the presynaptic input follows spiking of the postsynaptic neuron (anti-causal). However, it has not been possible to study the long-term consequences of such sequences of activity due to the invasive nature of the intracellular recordings that are needed to assess synaptic strength and control spiking of the postsynaptic neurons. We have developed an optogenetic method for studying spike-timing-dependent plasticity (oSTDP), thereby greatly extending the observation window from approximately 30 minutes to three days in brain (hippocampal) slice cultures. We use CheRiff and ChrimsonR to independently control pre- and postsynaptic spiking and observe changes in synaptic strength consistent with the â€˜classicalâ€™ spike-timing-dependent plasticity rules described above, during whole-cell patch-clamp recordings. Surprisingly, when the neurons are left untouched during oSTDP induction and relative synaptic strength is assessed 3 days later, only potentiation is observed regardless of the pre-post timing sequence. We have also developed an extremely violet/blue light sensitive potassium channel that can be combined with a Chrimson variant to switch neurons on or off with yellow/red or violet light, respectively. Using these tools we are exploring further the rules that set synaptic strength and that determine the pattern of connections in the hippocampal circuit.

Optogenetic Entrainment of Slow Brain Rhythms in Alzheimers Mice

Conference: Optogenetic Technologies and Applications

Year: 2019

Proceeding: Optogenetic Technologies and Applications

Group: General Submissions

Session: Optogenetics in Biomedicine - Neurobiology

Authors:

Kastanenka, K., MassGeneral Institute of Neurodegenerative Diseases

Alzheimerâ€™s disease (AD) is a major neurogenerative disorder without a cure. Slow brain rhythms, including slow oscillations, are important for consolidation of memory during sleep, and AD patients experience memory disturbances. Thus, we sought to examine slow wave activity using the voltage-sensitive dye RH1691 in an animal model of AD (APP mice). The power of slow oscillations at 0.6Hz was decreased starting at 3 months of age. Soluble amyloid-beta was sufficient to disrupt the slow waves. Next we sought to manipulate slow waves in APP mice with optogenetics. Driving slow oscillations at normal frequency with light activation of channelrhodopsin-2 (ChR2) restored slow wave power by synchronizing neuronal activity. Using multiphoton microscopy, we performed longitudinal imaging of senile plaques and monitored intracellular calcium. Cytosolic calcium is a surrogate marker of neuronal activity and is normally tightly regulated. We had previously demonstrated that resting calcium levels measured with the genetically encoded calcium sensor YC3.6 were elevated in a subset of neurons in APP transgenics, and hypothesized that an effective treatment would restore calcium to control levels. Driving slow oscillation activity with optogenetics halted amyloid plaque deposition and prevented calcium overload associated with this pathology. On the other hand, driving slow oscillation activity at twice the normal frequency (1.2Hz) resulted in increased amyloid production, increased amyloid plaque deposition, disruptions in neuronal calcium homeostasis, and loss of synaptic spines. Therefore, use of optogenetics allowed to determine that while restoration of physiological circuit dynamics was sufficient to abrogate the progression of Alzheimer's disease pathology and should be considered an avenue for clinical treatment of patients with sleep disorders, pathophysiological stimulation of neuronal circuits leads to activity dependent acceleration of amyloid production, aggregation and downstream neuronal dysfunction.

Optogenetics: Tools for Analyzing and Controlling the Brain

Optogenetic
Technologies & Applications

Conference: Optogenetic Technologies and Applications

Year: 2019

Proceeding: Optogenetic Technologies and Applications

Group: General Submissions

Session: Optogenetics in Biomedicine - Neurobiology

Authors:

Boyden, E. S., Massachusetts Institute of Technology

Understanding and repairing complex biological systems, such as the brain, requires new technologies that enable such systems to be observed and controlled with great precision, across extended spatial and temporal scales. We are discovering new molecular principles that are leading to such technologies. For example, we recently discovered that it was possible to physically magnify biological specimens manyfold, in an even way, by embedding them in dense swellable polymers, mechanically homogenizing the specimens, and then adding water to isotropically swell the specimens. In this method, which we call expansion microscopy (ExM), we enable scalable, inexpensive diffraction-limited microscopes to do large-volume nanoscopy, in a multiplexed fashion â€“ important, for example, for brain mapping. As another example, we discovered that microbial opsins, genetically expressed in neurons, could enable their electrical activities to be precisely driven or silenced in response to millisecond timescale pulses of light. These tools, called optogenetic tools, are enabling causal assessment of the contribution of defined neurons to behaviors and pathologies in a wide variety of basic science settings. Finally, we have developed new methods of directed evolution, and discovered mutant forms of optogenetic tools that enable precision fluorescent imaging of the high-speed voltage of neurons in the living brain. We share all these tools freely, and aim to integrate the use of these tools so as to lead to comprehensive understandings of neural circuits.

Regulation of Signaling Proteins in the Brain By Light

Conference: Optogenetic Technologies and Applications

Year: 2019

Proceeding: Optogenetic Technologies and Applications

Group: General Submissions

Session: Optogenetics in Biomedicine - Neurobiology

Authors:

Lamprecht, R., University of Haifa

In order to study the role of signaling proteins, such as kinases and GTPases, in brain functions it is necessary to control their activity at the appropriate spatiotemporal resolution and to examine the cellular and behavioral effects of such changes in activity. Reduced spatiotemporal resolution in the regulation of these proteins activity will impede the ability to understand the proteins normal functions as longer modification of their activity in non-normal locations could lead to effects different from their natural functions. We use and develop photoactivatable proteins to control signaling proteins, such as receptor tyrosine kinases and small GTPases, at high spatiotemporal resolution in neurons in behaving animals. For example, we show that photoactivation of EphB2, a key brain receptor tyrosine kinase, in pyramidal neurons in lateral amygdala during, but not after, learning leads to enhancement of long- but not short-term memory formation (e.g. Alapin et al., 2018). Activation by light of the small GTPase Rac1 (photoactivatable Rac1) in lateral amygdala during learning impaired long-term memory consolidation (e.g. Das et al., 2017). We currently further utilize and develop new optogenetic tools to control intracellular proteins (e.g. actin cytoskeleton regulatory proteins) to elucidate the cellular and molecular mechanisms in brain underlying behavior and memory formation.

Wireless optoelectronic systems for closed-loop regulation of pain and organ dysfunction

Optogenetic
Technologies & Applications

Conference: Optogenetic Technologies and Applications

Year: 2019

Proceeding: Optogenetic Technologies and Applications

Group: General Submissions

Session: Optogenetics in Biomedicine - Neurobiology

Authors:

Gereau, R., Washington University

The ever-expanding set of optogenetic tools has provided many elegant approaches for manipulating activity and signaling in cells, and has led to an explosion of new findings in neuroscience. Widespread implementation of optogenetics in studies of the peripheral nervous system (PNS), however, has been slowed by a unique set of obstacles, notably the challenge of consistent and restricted delivery of light to the neurons expressing opsins in the PNS. This presentation describes the development of an evolving set of tools for wireless, targeted illumination of neurons in the peripheral nervous system. As a case study, we present studies describing the implementation of wireless optogenetic manipulation of sensory afferents innervating the bladder for the control of bladder sensation and function. The devices and real-time analytical capabilities developed enable closed-loop manipulation of bladder pain and voiding dysfunction in freely moving animals, and suggest the potential for broad application of closed-loop manipulations to regulate a variety of physiological processes.

Chr-Based Optogenetic Approaches for Vision Restoration

Optogenetic Technologies & Applications

Conference: Optogenetic Technologies and Applications

Year: 2019

Proceeding: Optogenetic Technologies and Applications

Group: General Submissions

Session: Optogenetics in Biomedicine - Other biomedical applications

Authors:

Pan, Z. H., Wayne State University School of Medicine

The loss of photoreceptor cells in individuals with retinal degenerative diseases leads to blindness. Currently, there is no effective treatment to restore vision to the blind. Optogenetics is a promising strategy for restoring vision through the ectopic expression of genetically encoded light sensors, particularly channelrhodopsins (ChRs), in surviving inner retinal neurons to impart light sensitivity to the retina. Since our first demonstration of the feasibility of this strategy by expressing ChR2 in retinal ganglion cells in blind mice, optogenetic vision restoration has been reported by using a variety of optogenetic tools and different retinal cell targeting strategies. Optogenetic therapy for treating blindness caused by RP is currently in clinical trials. In this talk, I will briefly review the current state of the field and discuss our recent and ongoing studies for developing better ChR tools and effective treatment strategies for vision restoration.

Engineered Chimeric Antigen Receptors for Photo-Tunable Remote Control of CAR T-Cells

Optogenetic
Technologies & Applications

Conference: Optogenetic Technologies and Applications

Year: 2019

Proceeding: Optogenetic Technologies and Applications

Group: General Submissions

Session: Optogenetics in Biomedicine - Other biomedical applications

Authors:

Nguyen, N., Institute of Biosciences and Technology, Texas A&M University
Huang, K., University of Massachusetts Medical School
Huang, Y., Institute of Biosciences and Technology, Texas A&M University
Han, G., University of Massachusetts Medical School
Zhou, Y., Texas A&M University

Chimeric antigen receptor (CAR) T-cell immunotherapy has demonstrated high potential for elimination of tumors, particularly in patients with CD19-positive lymphoma and leukemia. CARs are synthetic receptors engineered onto the surface of T cells, where they can engage specific tumor antigens in a major histocompatibility complex (MHC)-independent manner. The recognition of antigen allows T cells to be activated and subsequently perform their killing/effector activities toward tumor cells. Despite the tremendous success of CAR T-cell therapy in cancer treatment, this type of immunotherapy imposes significant safety challenges, as most notably exemplified by the cytokine release syndrome (CRS) and the â€œon-target, off-tumorâ€ cytotoxicity due to the lack of precise control over the location and duration of anti-tumor immune response. Herein we present the design of light-switchable CAR T-cell (designated â€œLiCARâ€) that enables photo-tunable activation of therapeutic T cells to induce tumor cell killing both*in vitro*and *in vivo*. When coupled with imaging guided surgically removable upconversion nanoplates that have enhanced near infrared (NIR)-to-blue upconversion luminescence as miniatured deep tissue transducers, LiCAR enables precise spatiotemporal control over CAR T cell-mediated anti-tumor therapeutic activity. This remotely controllable nano-optogenetic device sets the stage for the exploration of optogenetic immunotherapy to deliver personalized anti-cancer therapy.

Engineering a Far-Red Light-Controlled Split-Cas9 Device for Genome Editing in Human Cells and Mice

Optogenetic
Technologies & Applications

Conference: Optogenetic Technologies and Applications

Year: 2019

Proceeding: Optogenetic Technologies and Applications

Group: General Submissions

Session: Optogenetics in Biomedicine - Other biomedical applications

Authors:

Yu, Y., East China Normal University
Ye, H., East China Normal University

The CRISPR-Cas9 technology has emerged as a powerful tool for targeted genome editing. However, uncontrollable expression of Cas9 greatly increases side effects due to the off-target cleavage. In order to address this problem and achieve conditional genome editing, several chemical inducible Cas9 systems have been developed. In this work, we demonstrated a far-red light-controlled split-Cas9 (FRL-Cas9) device that enables spatiotemporal control of genome editing in human cells and in a tdTomato reporter mouse model. In the meanwhile, we also showed that this FRL-Cas9 device could also be used to knock out oncogenes in tumor-bearing mouse model for cancer therapy simply by illumination. This optogenetically controlled device demonstrated a robust genome editing performance that exhibits remarkable reversibility, non-invasibility, negligible phototoxicity and deep tissue penetration for in vivo applications. This FRL-Cas9 device enables optogenetic control of targeted genome editing, which facilities better understanding of complex gene networks in basic biological research. Moreover, this optogenetic device opens new opportunities for dynamic and spatiotemporal interventions in gene-based precision medicine.

Optogenetic pharmacology of neuromodulatory receptors

Conference: Optogenetic Technologies and Applications

Year: 2019

Proceeding: Optogenetic Technologies and Applications

Group: General Submissions

Session: Optogenetics in Biomedicine - Other biomedical applications

Authors:

Isacoff, E., University of California Berkeley

A major goal of neuroscience is to understand how neuromodulatory systems regulate core processes of brain and behavior, from motor function and learning to reward, aversion, attention, and sleep. These systems go awry in schizophrenia and disorders of mood, motor control and cognition. Treatment for these conditions often turns to pharmacological manipulation of neuromodulators and their receptors. Understanding of neuromodulatory circuits has advanced considerably thanks to optogenetics and chemogenetics. But neuromodulation is difficult to crack. A major obstacle is that a single neuromodulator may play many diverse roles (e.g. dopamine functions in motor control, reward, aversion, attention, appetite and more). There are several reasons for this complexity: i) each neuromodulator has multiple receptor subtypes, and these may couple to distinct G proteins, and, even when coupled to the same G protein, may activate distinct effectors, ii) a particular receptor may have different functions in different cell locations (e.g. axons *versus* dendrites, and different micro-domain signaling complexes), and iii) a particular receptor may have different functions in different cells within a circuit.

We will describe a new approach that overcomes these limitations by controlling native, full-length neuromodulatory receptors at their natural cellular locations with photoswitchable orthogonal remote tethered ligands (PORTLs): a breakthrough of precise, receptor-, cell type- and brain region-selective switchable pharmacology.

A Yeast Optogenetic Toolkit for Control of Intra- and Intercellular Signaling

Conference: Optogenetic Technologies and Applications

Year: 2019

Proceeding: Optogenetic Technologies and Applications

Group: General Submissions

Session: Optogenetics in Biotechnology - Microbes

Authors:

McClean, M., University of Wisconsin-Madison
Sweeney, K., University of Wisconsin-Madison
Geller, S., University of Wisconsin-Madison
Moreno Morales, N., University of Wisconsin-Madison
An-adirekkun, J., University of Wisconsin-Madison

Optogenetic proteinsâ€”light sensitive proteins that change activity in their photoexcited stateâ€”make it possible to harness the exquisite spatiotemporal control available with light to actuate cellular activities. We describe a yeast optogenetic toolkit (yOTK) that allows for rapid assembly of optically controlled biological circuits within the model organism *Saccharomyces cerevisiae.*

We created an optogenetic tools capable of regulating gene expression and subsequent protein expression in *Saccharomyces cerevisiae* in response to blue light. A CRY2/CIB1based optogenetic activator increased expression of a target gene in response to blue light in a time- and dose-dependent manner. Additionally, a flexible optogenetic tool targeted to DNA through a nuclease-dead Cas9 protein was able to repress expression of target genes in a guide-RNA dependent manner. Our optogenetic tools are designed as DNA â€œpartsâ€ so that optically tunable circuits can be rapidly created using a modular cloning (â€œMoCloâ€) strategy. The modular nature of the toolkit allows us to design, construct, verify, and test optically sensitive circuits rapidly, with the entire cycle taking approximately two weeks.

Using the toolkit we have designed synthetic transcription factors with light-controllable nuclear localization and activity. By controlling transcription factor dynamics, we hope to uncover relationships between transcription factor activity, promoter structure, and gene expression. Furthermore, we can control intracellular interactions between cells using light. The spatial control available with light allows us to create structured populations of yeast, for example, through manipulation of expression of a public good.

Application of an Optogenetic Switch for Controlling the Central Carbon Metabolism in Escherichia coli

Optogenetic
Technologies & Applications

Conference: Optogenetic Technologies and Applications

Year: 2019

Proceeding: Optogenetic Technologies and Applications

Group: General Submissions

Session: Optogenetics in Biotechnology - Microbes

Authors:

Toya, Y., Osaka University
Shimizu, H., Osaka University
Tandar, S. T., Osaka University
Senoo, S., Osaka University

Optimization of metabolic flux distribution on pathways is an important approach for efficient bio-production using microorganisms. In particular, dynamic controlling the metabolic flux is a promising way for enhancing the productivity. To adjust the flux distribution, fine-tuning the expression level of metabolic enzymes is highly required, and inducible systems using chemicals have been frequently used. However, such chemically-inducible systems cannot be applied as a dynamic on/off control of metabolic pathways, because it is difficult to immediately remove added chemicals from the culture. Optogenetic switch is an attractive technique for controlling gene expression levels and for realizing a dynamic bio-process control, because light can readily be applied and removed from the culture multiple times.

In the present study, we applied a CcaS/CcaR optogenetic system to control a flux ratio of glycolytic pathways such as Embdenâ€“Meyerhofâ€“Parnas (EMP) and oxidative pentose phosphate (OPP) pathways in*Escherichia coli*. The expression of phosphoglucose isomerase (Pgi) that pulls glucose-6-phosphate at the branch point into the EMP pathway was regulated. The output EMP flux from the switch was optimized by adjusting plasmid copy number and by tagging Pgi with ssRA protein degradation signal. We evaluated the light-induced redirection of glycolytic flux with ^{13}C-metabolic flux analysis. The EMP:OPP flux ratio was controlled to 50:49 and 0.5:99 by exposure to green and red light, respectively [1]. Furthermore, we also demonstrated the CcaS/CcaR optogenetic switch for controlling the carbon flow between the EMP and methylglyoxal pathway in *E. coli* [2]. These switches would be useful tools for metabolic engineering.

[1] Tandar et al. Optogenetic switch for controlling the central metabolic flux of *Escherichia coli*. Metab Eng. 2019; 55:68-75.

[2] Senoo et al. Light-inducible flux control of triosephosphate isomerase on glycolysis in *Escherichia coli*. Biotechnol Bioeng. In press.

Optical Control of Exopolysaccharide Production and Biofilm Formation in the Soil Bacterium Sinorhizobium Meliloti

Optogenetic
Technologies & Applications

Conference: Optogenetic Technologies and Applications

Year: 2019

Proceeding: Optogenetic Technologies and Applications

Group: General Submissions

Session: Optogenetics in Biotechnology - Microbes

Authors:

Pirhanov, A., University of Connecticut
Cho, Y., University of Connecticut
Shor, L. M., University of Connecticut
Gage, D. J., University of Connecticut

â€œOptical control of exopolysaccharide production and biofilm formation in the soil bacterium *Sinorhizobium meliloti* â€œ

The gram-negative soil bacterium *Sinorhizobium meliloti* fixes nitrogen in symbiotic association with alfalfa plants. Establishment of this symbiosis depends upon the exchange of small signaling molecules between the soil bacterium and its host. In particular, exopolysaccharides produced by *S. meliloti* are required for invasion of root nodules, through a process that is not well understood. There are two types of exopolysaccharides produced by *S. meliloti*: succinoglycans (also called EPS-I) and galactoglucans (also called EPS-II). Besides aiding plant root association, exopolysaccharides are also thought to play a key role in protecting rhizobium cells against various environmental stresses, such as desiccation. In addition to its signaling and protective function, exopolysaccharides, in particular EPS-II, play a critical role in the development of structured complex 3-D surface attached communities of microbial biofilms in soil. Thus, in-depth understanding of properties and functions of exopolysaccharides secreted by *S. meliloti* cells is critical for the development of next generation technologies. While studying the interactions of exopolysaccharides with soil particles and its effect on moisture retention, we needed a tool that would allow us spatiotemporally control the production of exopolysaccharides with high precision. Ideally, a tool that allows on demand control of exopolysaccharide production, irrespective of type of media used and surface exposed, with minimal intervention would be desirable for this purpose. Here we demonstrate the development of such tool that allows us to spatially control the production of EPS-II and biofilm morphology, with ~400 μm resolution, in *S. meliloti* by using previously reported light responsive DNA binding protein EL222.

Optogenetic control of gut bacterial metabolism

Conference: Optogenetic Technologies and Applications

Year: 2019

Proceeding: Optogenetic Technologies and Applications

Group: General Submissions

Session: Optogenetics in Biotechnology - Microbes

Authors:

Tabor, J., Rice University

Gut bacteria secrete metabolites that impact the host. However, studies of the molecular mechanisms by which the microbiome impacts the host are challenging due to the poor accessibility of the gut environment. Here, we demonstrate that optogenetics can be used to remotely control Escherichia coli gene expression and metabolism in the gastrointestinal tract of live Caenorhabditis elegans worms. Using this approach, we reveal that gut bacterial secretion of the exopolysaccharide colanic acid (CA) protects worm intestinal mitochondria from stress-induced fragmentation. In addition, we show that light-activated CA secretion extends worm lifespan. Optogenetic control of gut bacteria will enable new mechanistic studies of how the microbiome impacts health and disease.

Optogenetics for Temporal and Spatial Control of Microbial Metabolism to Improve Chemical Production

Conference: Optogenetic Technologies and Applications

Year: 2019

Proceeding: Optogenetic Technologies and Applications

Group: General Submissions

Session: Optogenetics in Biotechnology - Microbes

Authors:

Avalos, J. L., Princeton University

The tunability and reversibility of light signals makes optogenetics a powerful strategy to dynamically and spatially control engineered microbial metabolisms for chemical production. My group has developed transcriptional optogenetic circuits to temporally control the expression of metabolic enzymes during the growth and production phases of microbial fermentations. These circuits can be designed to invert or amplify metabolic responses to light inputs to boost the production of various chemicals. The activation rates of these circuits can have a significant impact on chemical yields and titers, and on our strategies to use open- or closed-loop controls of metabolism; however, we have found ways to manipulate these activation rates to our convenience. To further increase the rates of response to light inputs beyond the limitations imposed by transcription, my group has also developed optogenetic controls that work at post-translational levels. Some of these controls are based on light-dependent assembly of synthetic membraneless organelles, which can be used to co-localize or disperse metabolic enzymes in subcellular space to control flux through branched metabolic pathways. Other systems are based on highly specific protein binders that can be used to control enzymatic activity posttranslationally, and can be engineered to be light-switchable. I will discuss how these various optogenetic technologies can be integrated to optimize microbial fermentations for chemical production.

Orthogonal Optogenetic Circuits for Optimizing Microbial Chemical Production

Conference: Optogenetic Technologies and Applications

Year: 2019

Proceeding: Optogenetic Technologies and Applications

Group: General Submissions

Session: Optogenetics in Biotechnology - Microbes

Authors:

Lalwani, M. A., Princeton University
Avalos, J. L., Princeton University

Microbial chemical production offers a sustainable alternative to traditional petrochemical processing. To circumvent metabolic burden resulting from expression of heterogeneous pathways, metabolic engineers implement dynamic control: accumulating sufficient amounts of biocatalyst in a growth phase, then activating metabolic pathways in a production phase. Optogenetic control of metabolic pathways presents an advantageous method for dynamic regulation of microbial metabolism: light is cheap, reversible, and tunable via pulsing in different duty cycles.

Our group has developed various optogenetic circuits that allow for flexible switching between cellular growth and chemical production. By optogenetically regulating the abundances of transcriptional activators and repressors, these circuits are able to induce gene expression in either blue light or darkness. Moreover, by implementing protein pairs that do not interfere with each otherâ€™s activities, we have allowed for simultaneous signal amplification and inversion. Our circuits display high expression levels, rapid activation kinetics, and minimal cross-talk. Through co-utilization of these optogenetic amplifier and inverter circuits, we are able to optimize production titers as well as tunably produce chemicals in different ratios. We believe that our results validate the potential of optogenetics for sustainable and economically feasible production of valuable chemicals.

Highlighter - a Synthetic Light-Gated Gene Expression Switch for Plants

Conference: Optogenetic Technologies and Applications

Year: 2019

Proceeding: Optogenetic Technologies and Applications

Group: General Submissions

Session: Optogenetics in Biotechnology - Plants

Authors:

Larsen, B., The Sainsbury Laboratory Cambridge University
Jones, A., The Sainsbury Laboratory Cambridge University

In the process of applying optogenetic sensors in plants to study phytohormone patterning it has become clear that high-resolution sensing alone is not sufficient - we also need to be able to perturb phytohormone patterning in high-resolution to distinguish correlation from causation. Optogenetic actuators are ideal tools for performing high-resolution perturbations in a minimally invasive manner with maximal spatiotemporal resolution. However, a major challenge for designing optogenetic actuators for plants is that plants are conventionally grown in mixed light environments, cycling between changing broad spectrum light regimes and darkness. Consequently, an optogenetic actuator for plants would ideally adopt one target gene expression state during light-dark cycling and an altered target gene expression state following treatment with light spectra that are not found in the horticultural environment.We have now developed an orthogonal synthetic light-gated gene expression system for plants which is fully genetically encoded and functions without the need for exogenously supplied chromophores. Our system, called â€˜Highlighterâ€™, was developed by repurposing the cyanobacterial CcaS-CcaR system through semi-rational engineering. When tested by transiently transfecting *N. benthamiana*, the system was inactive under blue light conditions and active in the dark and under green, red and white light conditions. Furthermore, Highlighterâ€™s activity could be tuned by modulating the red-to-blue ratio of white light. We are presently adapting Highlighter for use in stably transformed plants and continue to optimize the system, engineer new spectral variants and envision using the system to perturb hormone dynamics in single cells of stably transformed plants with rapid induction and reversion kinetics using minimally invasive light signals.

Optogenetic Manipulation of Guard Cell Ion Transport Enhances Plant Water Use Efficiency and Carbon Capture

Conference: Optogenetic Technologies and Applications

Year: 2019

Proceeding: Optogenetic Technologies and Applications

Group: General Submissions

Session: Optogenetics in Biotechnology - Plants

Authors:

Blatt, M., University of Glasgow
Papanatsiou, M., University of Glasgow
Christie, J., University of Glasgow
Wang, Y., Zheijiang University
Petersen, J., University of Jena

Plant growth and productivity are strongly correlated with water availability that has been a limiting factor, especially at arable lands. The demand for engineering crops to withstand such extreme conditions without jeopardising growth and yield is higher than ever. The control of ion transport in guard cells dictates stomatal movements enabling the uptake of CO_2 in expense of water vapour release, and therefore optimizing the trade-off between transpirational water loss and carbon gain. Here, we showcase the potential of optogenetic engineering to modulate guard cell ion transport in order to improve water use efficiency. Blue light activation of blue-light gated K^+ channel (BLINK) was confirmed in mutants impaired in stomatal opening under blue light. We further validated the photoactivation of BLINK by showing the BLINK-expressing lines having higher K^+ conductance in blue light. Gas exchange measurements revealed faster stomatal movements and improved water use of BLINK-expressing plants when grown under fluctuating light condition, and in combination with water stress. We therefore propose the use of BLINK as an optogenetic tool to alter stomatal behaviour and improve water use efficiency without a penalty in carbon fixation.

AUTHOR INDEX